A Night in Fernwood Forest

Written by Brenda Boreham
Illustrated by Judith Rackham

A Night in Fernwood Forest

Written by Brenda Boreham

Illustrated by Judith Rackham

ISBN: 978-1-77174-602-1

Visit: www.strongnations.com

Contact: 1-888-278-2202

Photo Credit

Page 30 - Don Henise: Ocean Spray Fairy Moth (Adela septentrionella)

Cataloging-in-Publication Data

Title: A night in Fernwood forest / written by Brenda Boreham; illustrated by Judith Rackham.

Names: Boreham, Brenda, author. | Rackham, Judith, illustrator. | Strong Nations Publishing, publisher.

Description: Nanaimo, British Columbia : Strong Nations Publishing, 2023. | 32 pages : colour illustrations; 8.5 x 11 inches. | By Day and By Night Nature Series; 4

Summary: "A Night in Fernwood Forest is about a walk along a secluded forest trail that begins at dusk and ends at dawn on a quiet night in May. Here we catch glimpses of ten animals as they move about in their woodland home. A mother cougar and her kitten, a colony of bats, and a family of shrews are among the nighttime creatures we encounter."-- Provided by publisher.

Identifiers: ISBN 9781771746021 | ISBN 9781771746014

Subjects: LCSH: Rain forests--Juvenile fiction. | LCSH: Rain forest animals--Juvenile fiction.

Classification: LCC PZ7 1 B674 N556 2023 | DDC jC813/.6-dc23

BRITISH COLUMBIA ARTS COUNCIL | BRITISH COLUMBIA

Supported by the Province of British Columbia

FSC www.fsc.org MIX Paper from responsible sources FSC® C016245

Strong Nations Publishing Inc.

Proudly printed in Canada

To Diana, Dianne, Judith, Lauren, Patrick, and Ron.
Thank you for your expertise, flexibility, patience, and support.
I am grateful to have such a great book-making team!

Brenda

The illustrations that capture
the magic of Fernwood Forest are dedicated to my
daughter, Eliza, and my granddaughter, Celeste.

Judith

The forest fills with the glow of the setting sun. In this forever moment, nothing stirs. The only sound is the rustle of a gentle breeze shifting through the treetops.

The sun hesitates for a moment and then sinks quietly out of sight. The moment passes. It is dusk.

The sun has set. The western sky fills with soft streaks of pink, apricot, and gold. The moon sits high in the eastern sky.

Silently, a barred owl lifts from his perch in an old maple tree. With wings outspread, he circles above the forest trail. He watches carefully for any movement in the litter of dry leaves and debris on the forest floor. Mice, squirrels, shrews, frogs, and other small animals are among the prey he is seeking.

It is evening. The warm glow of sunset has faded, and the light is dimming. A sprinkle of stars appears in the clear night sky. Moonlight filters softly through the tree branches to the forest floor.

With a twitch of her long, black-tipped tail, a mother cougar pads quietly past the old maple tree. She heads down the trail with her cub following closely at her side. They turn off the trail, cross a glade, and wind through the trees to drink at a small pond. They lap up some water, then turn and slip silently back into the shadows.

It is midevening. Stars now fill the sky! Silver moonlight washes across the forest and reflects in the pond on its western edge.

The pond sits quiet and still in the darkness. Suddenly, a boisterous “ribbit!” breaks the silence. A male Pacific tree frog calls loudly from the shallow water at the edge of the pond. A second frog, then a third, joins in the chorus. The pond is soon a riot of sound!

By late evening, the stars glitter brightly across the darkened sky. The moon continues on its westward path.

With the light of the moon guiding them, a family of raccoons leaves the pond. They scramble through the trees and run across the glade to a thick patch of brambles. Growing among the prickly stems and clusters of dark green leaves are some juicy, ripe salmonberries! The family chatters happily as they begin to feast.

It is midnight. The stars gleam overhead as the moon shifts slowly to the west.

A small, grey deer mouse crouches quietly in the darkness. With her sharp teeth, she gnaws through the hard coat of a seed she had hidden under the ferns. After finishing her meal, she scurries to her nest beneath the root of an old fir tree, where her children are curled up asleep.

The starlight shines steadily on the pond and forest below. The moon dips toward the treetops.

With a flutter of purplish-brown wings, moths circle around the fir tree. Their long antennae sniff the air for sap leaking from scars on the broken branches. They are hoping for a taste of the sap or of nectar from night-scented flowers.

The moon slowly sinks out of sight behind the trees. The forest floor lies in shadow—only a faint glow of light from the starry sky filters through the leaves and branches.

A large, brown mosquito sits resting with her head down and tail up on the leaf of a trillium plant. Full of sweet nectar, she spreads her dark-spotted wings and takes flight. She rejoins a swarm of mosquitoes hovering around an old stump.

The stars gleam overhead as the night passes. The moon drops below the horizon.

With dawn approaching, a colony of little brown bats is on its last hunting trip of the night. The bats find small flying insects by calling out and listening for an echo as the sound bounces off objects around them. A feast of beetles, mosquitoes, flies, moths, and midges is a fine meal before they rest for the day.

In the last hour of darkness, the stars flicker brightly above the forest canopy.

Heaps of fallen leaves, dead twigs, and branches lie around the base of the old maple tree. Hidden under the tangle of debris, a vagrant shrew rests with her family. Their cup-shaped nest, lined with soft hair, grass, and leaves, has a leafy, domed roof to protect the family from the rain.

The last few moments of darkness come to an end.

Pacific sideband snails are well-camouflaged in the piles of leaf litter at the base of the maple tree. The snails move slowly over the damp earth, safely hidden from the eyes of mice, snakes, raccoons, and shrews. They stop here and there to chew on leaves, fungi, moss, and other bits of decaying vegetation.

The forest lies peacefully under a blanket of fading stars. In this forever moment, not a creature stirs. The only sound is the rustle of a gentle breeze shifting through the treetops.

A soft, grey light begins to filter through the leafy, green canopy. The moment passes. It is dawn.

About the Pacific Northwest Temperate Rainforest

What plants grow in the Pacific Northwest temperate rainforest?

Trees, bushes, mosses, and ferns are among the many plants that grow in the Pacific Northwest temperate rainforest. Plants are living things. They grow, take water and nutrients from the soil, and reproduce. Scientists divide plants into two groups: those that reproduce from seeds and those that reproduce from spores.

Plants from Seeds

Most plants reproduce from seeds. Each plant species makes seeds distinct in size and shape. Each seed contains a tiny plant and a supply of food stored inside its protective coat. The seeds grow into plants when they have just the right temperature and the right amount of water, light, and soil.

Trillium

Thimbleberry

Maple Tree

Douglas Fir

Plants from Spores

Some plants, as well as most fungi, reproduce from spores. Spores are single cells that grow when the soil, temperature, moisture level, and sunlight are just right.

Sword Fern

Moss

Horsetail

Bracken Fern

Barred Owl (pages 6, 7) The barred owl is a bird. It ranges in size from 40 to 60 centimetres (16 to 24 inches). It has brown and white striped markings on its body. It has a small, sharp, yellow beak and sharp talons. The barred owl is a carnivore. It hunts by day and night for small animals, such as squirrels, mice, rabbits, and frogs.

Cougar (pages 8, 9) The cougar is a mammal. It is one of the largest wild cats in North America. It can grow to 3 metres (10 feet) from its nose to the tip of its tail. It varies in colour from reddish-brown to greyish-brown or tan. It has lighter markings on its throat and chest. The cougar is a carnivore. It hunts deer, elk, and big-horned sheep, as well as smaller animals, such as rabbits.

Pacific Treefrog (pages 10, 11) The Pacific treefrog is an amphibian. It is a small frog that reaches about 5 centimetres (2 inches) in length. The Pacific treefrog's colour ranges from green to brown depending on the temperature, the available light, and its habitat. It is a carnivore. Flies, wasps, ants, spiders, and snails are some of the creatures it hunts.

Raccoon (pages 12, 13) The raccoon is a mammal. It ranges in size from 40 to 70 centimetres (16 to 28 inches) in length. It is grey to reddish-brown and has a black mask around its eyes. Its bushy tail has four to ten dark rings. The raccoon is an omnivore. It eats a wide variety of food, such as birds' eggs, mice, insects, and fruit.

Deer Mouse (pages 14, 15) The deer mouse is a small mammal that grows up to 10 centimetres (4 inches) in length. It has soft, grey fur on its back and usually has white fur on its lower body. It has a long tail, stiff whiskers, and black eyes. The deer mouse is an omnivore. It forages for food, such as seeds, nuts, insects, and snails.

Adela Septentrionella (pages 16, 17) The insect Adela septentrionella is a species of moth. It has long antennae and four wings, which are black with two white stripes. The length of its forewing is up to 5.7 millimetres (0.2 inches). Males have black heads; females have orange heads. Adela septentrionella is an herbivore. It eats tree sap or nectar from flowers.

Culiseta Mosquito (pages 18, 19) The Culiseta mosquito is an insect. It is a large mosquito that grows to 10 millimetres (0.4 inches) in length. It has dark spots on its wings. Adult mosquitoes feed on flower nectar. Females also feed on blood, which supplies nutrients required for reproduction.

Little Brown Bat (pages 20, 21) The little brown bat is a mammal. It ranges in size from 6 to 10 centimetres (2 to 4 inches). It has brown fur and dark brown or black wings. The little brown bat is a carnivore. It usually hunts right after dusk and just before dawn for moths, flies, beetles, mosquitoes, and other insects.

Vagrant Shrew (pages 22, 23) The vagrant shrew is a mammal. Its fur ranges in colour from grey to dark brown. It grows up to 10 centimetres (4 inches) in length and has a tail that measures 4 centimetres (1.6 inches). It is an omnivore. Shrews living on the west coast of North America eat slugs, flies, snails, seeds, and fungi. During the spring, this shrew is most active just after dusk and at dawn.

Pacific Sideband Snail (pages 24, 25) The Pacific sideband snail is a mollusk. It is the largest land snail in British Columbia, growing to 3.5 centimetres (1.4 inches) in diameter. Its shell ranges in colour from dark brown to pink and has a dark band around the perimeter. The Pacific sideband snail's body is pink with two tentacles on its forehead. It eats most kinds of vegetation.

Nurse Log

Nurse logs are trees that have fallen and are decaying on the forest floor. Over time, the wood gradually rots and is broken down by insects, plants, fungi, and the weather. A nurse log provides shelter and nourishment for many plants and animals as it decays.

Douglas Fir Sapling
Sphagnum Moss
Salal
Western Redback Salamander
Raccoon
Deer Mouse
Pacific Sideband Snails
Banana Slug
Pill Bug

Welcome to the Pacific Northwest Temperate Rainforest

The plants and animals of the Pacific Northwest temperate rainforest are amazing! They have learned to survive the changes in temperature, daylight hours, and the amount of precipitation that falls throughout the four seasons. Enjoy visiting their homes, and please show respect for all living things.

- The forest is home to many living things. We stay on designated trails to prevent damaging this delicate ecosystem.
- We pick things up only if we are sure they are not living (feathers, nuts, cones, stones, fallen leaves, etc.). We return these things to the forest floor before we go home.
- Plant collection is not permitted in any provincial or national parks. We leave the plants, seeds, and flowers undisturbed.
- We treat all creatures with respect. We observe, but we do not approach them or touch them. We keep our pets on a leash at all times.
- We check the B.C. Parks website for campfire bans or closures before we enter the forest.
- We have campfires only in designated places, and we follow the campground safety rules.
- If we spot a wildfire in British Columbia, we report it immediately: 1-800-663-5555.
- When it is time to leave the forest, we take away everything we brought with us.